curious about

SAVING EARTH

BY AMY S. HANSEN

AMICUS LEARNING

What are you

curious about?

Curious About is published by Amicus Learning, an imprint of Amicus.
P.O. Box 227
Mankato, MN 56002
www.amicuspublishing.us

Editor: Alissa Thielges
Series Designer: Kathleen Petelinsek
Book Designer and Photo Researcher: Emily Dietz

Library of Congress Cataloging-in-Publication Data
Names: Hansen, Amy, author.
Title: Curious about saving Earth / by Amy S. Hansen.
Description: Mankato, MN: Amicus Learning, [2025] | Series: Curious about green living | Includes bibliographical references and index. | Audience: Ages 5–9 | Audience: Grades 2–3 | Summary: "What can we do to save the Earth? Ignite kids' growing curiosity about the environment with an inquiry-based approach to saving the planet. Includes infographics and back matter to support research skills, plus table of contents, glossary, and index"—Provided by publisher.
Identifiers: LCCN 2023043272 (print) | LCCN 2023043273 (ebook) | ISBN 9781645496984 (library binding) | ISBN 9781681529677 (paperback) | ISBN 9781645497042 (ebook)
Subjects: LCSH: Environmental protection—Juvenile literature. | Climate change—Juvenile literature. | Plastic scrap—Environmental aspects—Juvenile literature. | Recycling (Waste, etc.)—Juvenile literature.
Classification: LCC TD170.15 .H356 2025 (print) | LCC TD170.15 (ebook) | DDC 363.7—dc23/eng/20231214
LC record available at https://lccn.loc.gov/2023043272
LC ebook record available at https://lccn.loc.gov/2023043273

Photo Credits: Getty: ArtMarie 16, Darryl Leniuk 9, izusek 10–11, Steve Clancy Photography 14–15; Noun Project: Estu Suhartono ID 22 & 23 (Earth), Gregor Cresnar 22 & 23 (tree); Shutterstock: AnnaStills 3, 20–21, badahos 19, Belozersky 7 (hurricane), BlueRingMedia 7, Ho Van Ty 18, Igor Iakovlev 6, Ingo Steinbach 2 (l), 4–5, KOTOIMAGES 17, Mohamed Abdulraheem 13, StockerThings cover, 1, structuresxx 2 (r), 12

Printed in China

CHAPTER ONE

Forests are often cleared to make space for farming.

Why does Earth need to be saved?

Our planet is in trouble. The **climate** is changing and getting warmer. The oceans have a lot of plastic. And **wilderness** is disappearing. These are some of Earth's biggest problems. But there is hope. By learning about what Earth needs, we can all help.

How is Earth getting warmer?

Power plants, cars, and forest fires put chemicals in the air. The chemicals trap heat on the planet. Earth is 2 degrees Fahrenheit (1 degree Celsius) warmer than it used to be. That may not seem like much. But it makes a BIG difference.

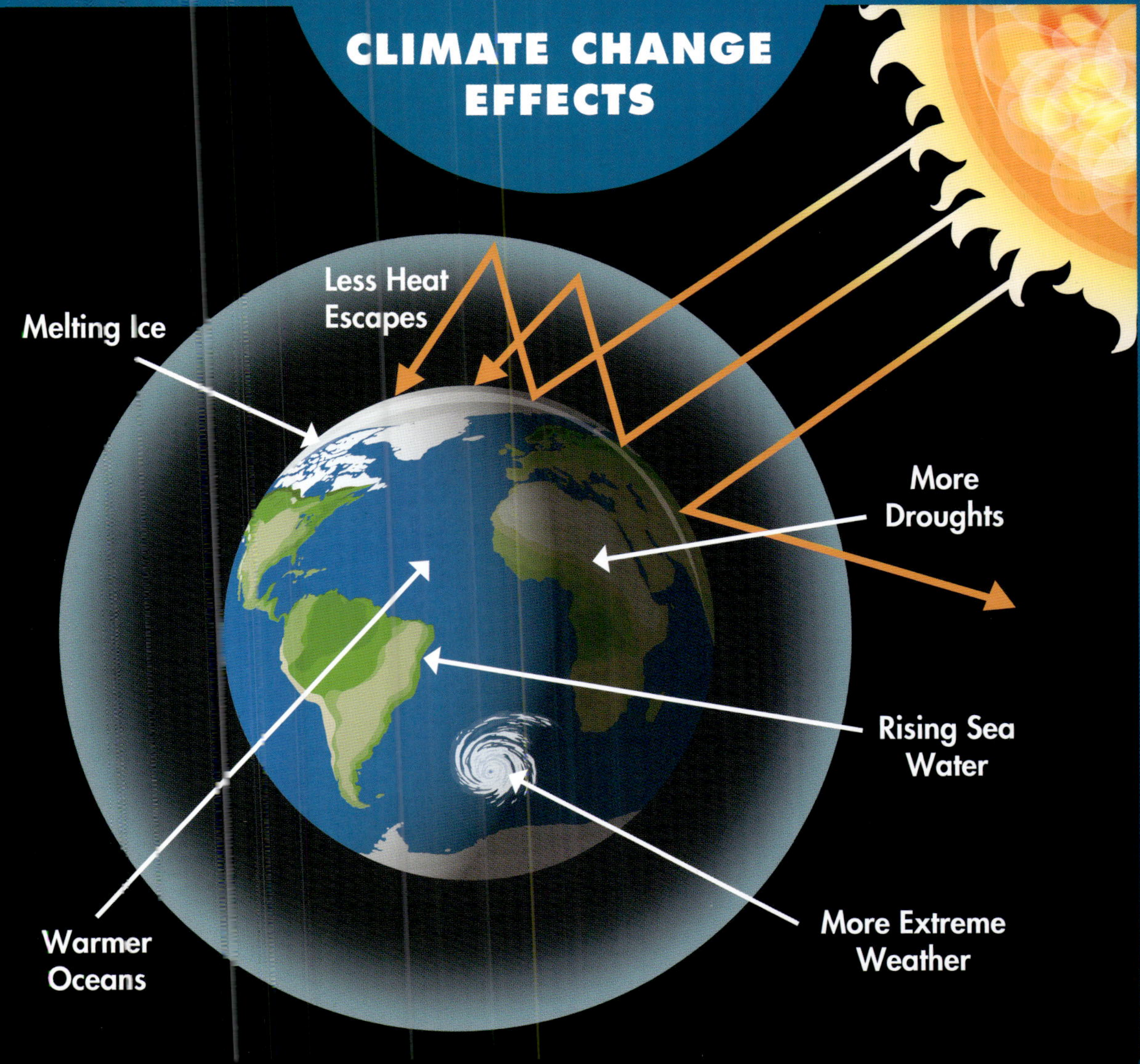

Is a warmer Earth bad?

Yes. It is bad for people's health. People need more water when it is hot. But fresh water is running out. Warm oceans hurt sea animals, too. Coral reefs are dying out. These are homes for many sea creatures. Warm temperatures also create more extreme storms.

AVERAGE WORLD TEMPERATURES

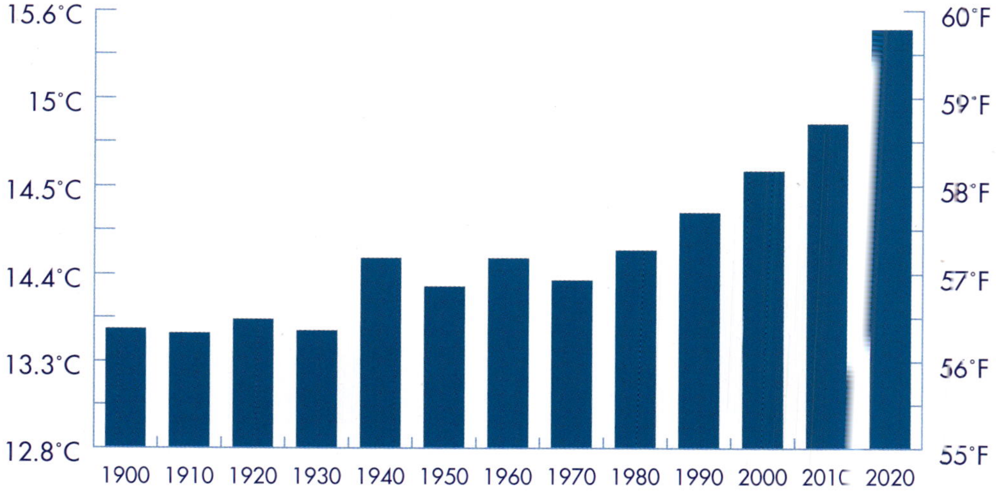

Coral reefs turn white and die when the ocean gets too warm.

How can I help Earth stay cool?

A family helps plant a new tree in a park.

Plant trees. Trees help Earth heal. They store **greenhouse gases** that harm the climate. Trees also give shade, which cools the ground. Green spaces like parks are cooler than buildings. By creating more green spaces with plants, we can cool big cities.

DID YOU KNOW?
In 2004, Wangari Maathai won the Nobel Peace Prize. She taught women in Kenya how to plant trees. Those women taught others. Together, they planted 30 million trees.

CHAPTER TWO

Animals can mistake plastic for food and get sick.

Why is plastic so bad?

DID YOU KNOW?
Plastic does not break down easily. A water bottle lasts 450 years in a landfill.

Plastic is hard to recycle. And it does not **decay**. Pieces can be found all over. People find plastic in water. Scientists have even found tiny pieces of **microplastic** in soil. Animals and humans can accidentally eat this plastic. This makes them sick.

How do plastics hurt Earth's climate?

Factories use **fossil fuels** to make plastic. The process puts chemicals in the air. These chemicals trap heat in the air. Drilling for fossil fuels also hurts the wilderness. Companies cut down trees and build roads so the oil can be moved. Oil spills can harm animals, too.

DID YOU KNOW?

The Great Pacific Garbage Patch floats near Hawaii. It is made of 1.8 trillion pieces of garbage that the ocean currents moved together. It is the size of the state of Alaska.

Oil washes onto New Zealand's shore after a spill in 2011.

What can I do?

A boy puts plastic bottles in a city recycling bin.

Use less plastic. And pick up garbage when you see it. Bring your own metal or wooden spoons with your lunch at school. Remind your parents to bring cloth bags to the store. Carry a refillable water bottle. And recycle when you can.

You can carry groceries in cloth bags.

CHAPTER THREE

Why is wilderness important?

Wilderness spaces help fight climate change. Plants and animals grow freely. The plants clean the air and water. Their roots hold soil in place during big storms. Animals **pollinate** plants and help move seeds. Wilderness areas help the whole world.

DID YOU KNOW?
Sunflowers help clean the soil. They can reduce pollutants. And they look pretty!

National parks protect wild spaces for animals and people to enjoy.

What can I do to help save the wilderness?

When you visit parks, be kind to the wilderness. Tell your friends and family why these places need to be protected. Do not leave a mess. Do not get close to animals. Offer to help clean litter from the trails.

A dad and son pick up trash in a park.

ASK MORE QUESTIONS

What does climate change mean for my state or province?

What is the longest wilderness trail in the United States?

Try a BIG QUESTION: All of Earth's problems are connected. Can you explain why plastics are a problem for the air, water and wilderness?

SEARCH FOR ANSWERS

Search the library catalog or the Internet.
A librarian, teacher, or parent can help you.

Using Keywords
Find the looking glass.

Keywords are the most important words in your question.

If you want to know about:

- how climate change is affecting your state, type: CLIMATE CHANGE [STATE NAME]
- the longest trail, type: LONGEST U.S. WILDERNESS TRAIL

LEARN MORE

FIND GOOD SOURCES

Are the sources good?
Some are better than others. An adult can help you. Here are some good, safe sources.

Books

Biodiversity by Anne O'Daley, 2023.

Healthy Planet by Elizabeth Carney, 2024.

Internet Sites

Kids vs. Plastic
https://kids.nationalgeographic.com/nature/kids-vs-plastic
National Geographic explores the planet and its problems. It is a good source for animals and nature.

Climate Action Superheroes
https://www.un.org/sustainabledevelopment/climate-action-superheroes/
The United Nations (UN) is made up of many countries who have pledged to help fight climate change.

Every effort has been made to ensure that these websites are appropriate for children. However, because of the nature of the Internet, it is impossible to guarantee that these sites will remain active indefinitely or that their contents will not be altered.

SHARE AND TAKE ACTION

Join a climate change group at your school.
If there is not a group already, create one! Your group can plant gardens and pick up litter.

Pack your lunch and snacks in reusable containers.
This will cut down on plastic bags. Instead of a juice box, bring a reusable water bottle.

Explore a state park.
Grab a buddy or two and go on a hike. Talk to a park ranger. You can ask them how they keep the wilderness clean for animals. Maybe someday you'll have that job!

GLOSSARY

climate The usual weather conditions in a particular place or region.

decay To slowly break down by natural means.

fossil fuel A material that is formed in the ground from dead plants or animals.

greenhouse gas A gas that traps heat in the atmosphere.

landfill An area where waste is buried under the ground.

microplastic Tiny pieces of plastic that harm the environment.

pollinate To move pollen from one plant to another; this allows the plants to grow new seeds.

pollutant A substance that makes land, water, or air dirty and not safe to use.

wilderness A wild and natural area where few people live.

INDEX

About the Author

Amy S. Hansen lives in Maryland. She started writing as a reporter for local newspapers. After a few years she went back to school to study environmental science and then worked as a science reporter. Nowadays, she (mostly) writes for kids because it is much more fun.